PEAT A[illegible]
PEAT CUTTING

Ian D. Rotherham

SHIRE PUBLICATIONS

Published by Shire Publications Ltd.
PO Box 883, Oxford, OX1 9PL, UK
PO Box 3985, New York, NY 10185-3985, USA
Email: shire@shirebooks.co.uk
www.shirebooks.co.uk

First published in 2009.
Transferred to digital print on demand 2016.

A CIP catalogue record for this book is available from the British Library.

Shire Library no. 483 • ISBN-13: 978 0 74780 705 6

Ian D. Rotherham has asserted his right under the Copyright, Designs and Patents Act, 1988, to be identified as the author of this book.

Designed by Ken Vail Graphic Design, Cambridge, UK.
Typeset in Perpetua and Gill Sans.
Printed and bound in Great Britain.

COVER IMAGE
A typical peat cut face showing the spade marks in the peat and the barrow used to cart the load away.

TITLE PAGE IMAGE
This Shetland lady is loading a pony's panniers with peat to carry to the cottage from the stack.

CONTENTS PAGE IMAGE
A typical old Irish farmyard with a variety of carts and a sled plus other equipment. Note the wicker peat basket as part of the sled.

DEDICATION AND ACKNOWLEDGEMENTS
This volume is dedicated to my wife Liz for putting up with an obsession, and to my friend and fellow enthusiast Paul Ardron for sharing a passion. Joan Butt helped with artwork and images.

All images are from the author's own collection.

CONTENTS

PREFACE

THIS BOOK presents an overview of peat, peatlands and their usage by people throughout history. It celebrates a cultural heritage from the Iron Age to the twentieth century, which gave us the Norfolk Broads and other distinct landscapes, and was a part of most people's lives. It also sets out to delve into some of the issues associated with this interesting and rich resource. My focus is on how communities lived in and used peat bog, fen, common and heath, taking a regional approach with examples from Ireland, Scotland (including the Scottish Islands), England and Wales. In some rural areas until the 1970s, every cottage had a peat stack almost as big as the building. The peat or turf fire, the hearth or the stove, had its own mystery and culture, the centre of household and daily life. Sadly, much of this is now lost and forgotten.

A peasant peat cutter with peat fork, and a very large peat stack.

Peat bogs and fens were a contested resource, so when Charles I tried to enclose and drain them it led to the Civil War and his execution. (This was a touch ironic as he had tried to foster the common rights of the ordinary people in the face of intransigent aristocratic ambitions.) However, with the demise of Charles, Oliver Cromwell, himself a fen man, finished the job of drainage and 'improvement'. The use of fen and bog was often essential for common

This image of an Irish hovel with turf roof illustrates the dire poverty found in many rural areas in the late nineteenth and early twentieth centuries.

survival but was challenged by landowners. Their complex administration and careful management reflected the importance of this resource, and ran continuously through the centuries to the great period of agricultural 'improvement' from 1700, even until the twentieth century. The enclosure of commons, with displacement of peasants and the poorer rural community, set the seal on many sites. The last great lowland fens were drained for arable farming during the war effort of the 1940s, traditions and folklore passing into distant memory.

The process of peat extraction also left imprints on local culture and heritage, but recent industrialisation of peat extraction has proved unsustainable and unacceptable. From the mid nineteenth century, there

Domestic peat cutting in Ireland: note the panniers and the sheep hitching a lift.

Shetland peat carriers.

Turfing on blanket bog with long-handled spades.

were many attempts to use peat in different ways, from making paper to home insulation and as a petrochemical substitute. In Europe and particularly Germany during the mid- to late nineteenth century, peat was used for peat baths and health spas. This was pioneered in Britain at Buxton and Harrogate in England, and at Strathpeffer in Scotland. For the right price you could be immersed in hot peat and have electric currents passed through you, or you might even opt for the hot rectal peat douche. Although popular in parts of Europe, in Britain these treatments lapsed by the late 1950s, and faded from

A massive industrial peat fuel press for making peat briquettes.

memory. Another important leisure association of peat is the making of malt whisky, illicit or otherwise – an aspect of peat bogs close to many hearts.

It should not be forgotten that there are huge conservation issues associated with peat and its exploitation. The destruction of peat bogs must rank as one of the greatest environmental disasters in Great Britain: loss of peat and peatlands contributed to climate change by releasing masses of carbon dioxide; and frequent flooding occurred owing to the removal of extensive areas of water-absorbent sponge. Peat and peatlands are in the news again with ambitious plans to reinstate some of the lost bogs and fens. This book rekindles memories of traditions and times now past; walking across many countryside areas may never be quite the same again.

INTRODUCTION

THE ROMAN WRITER Pliny in his *Natural History* described peat cutting in the first century, by German tribes along the Rivers Elbe and Ems: 'They weave nets of rushes and sedges to catch fish; and form mud with their hands, which, when dried in the wind rather than in the sun, is burned to cook their food, and warm their bodies chilled by the cold north wind.'

There is archaeological evidence for peat cutting in Denmark and in the Fens and Somerset Levels in the pre-Roman period. Peat-cutting tools, and even cut turves, have been identified that are over 2,000 years old. A Classical description of the Celtic Batavi tribe suggested that they were so wretched that '... their drink is the drink of swine, and they burn their very earth for warmth', meaning that they drank beer and burned turf. Early writers referred to peat as 'combustible earth', including Cardinal Piccolomini, who in 1458 described how the people of Friesland made '... fires of combustible earth, since they lacked firewood'.

The use of peat for fuel has almost ceased in most areas. Many people associate peat cutting with Ireland or the Western Isles of Scotland, or late twentieth-century industrial extraction. Yet historically, peat and turf were the common fuels for most people across the whole country. Indeed the mark of the peat cutter is written deep in the landscapes of Britain. In regions such as Somerset, Cornwall, and North Wales, older people still recall the distinctive 'peat reek' in their cottages. Associated with traditional community peat cutting and use were distinctive tools and implements, and long-standing cultural attachments and folklore. In Scotland and the Isles, and in Ireland itself, peat cutting is still important in culture, folk memory and identity. However, across much of Wales and most of England, the memories are long past. Towns and cities such as York, Norwich, Kendal, Carlisle, and even Liverpool, depended on peat fuel. The former moss-lands are written into today's landscapes as place-names, even when all other evidence has gone. In recent times peat was used as fuel in the Fens, in the Somerset Levels, in the Lake District and North Lancashire, on the North Yorkshire Moors, in Devon and Cornwall, and in North Wales. In almost all

Opposite:
Peat cutting in Scotland in the eighteenth century. Teams of men are working the peat and the cut turves are carried away in carts.

A typical cottage on Achill Island, Ireland, with its thatched roof tied down, and its turf stacks, the vital store of winter fuel.

A cottage with peat stack on the Isle of Skye. The crofter is collecting peats for the fire and the stack is part-used. Peat smoke can be seen wafting from the chimneys.

This long crofter's cottage on the Isle of Lewis has its peats and loaded peat cart pulled by a pony.

these there is no longer any active cutting. There remain perhaps three peat cutters in North Yorkshire; in West Yorkshire the Graveship of Holme is England's last organised community peat moss.

How peat was cut, processed and used varied through time and between regions – a rich tapestry in danger of loss from cultural memory. Local rural history museums preserve some implements and even buildings complete with traditional hearths. A deep sense of history and mystery concerns the traditions of common rights and their attachments to buildings or hearths. Famous cottage fires and particularly those in old inns burnt continuously for decades or even centuries. Local people placed huge importance on these.

Finally, from the mid 1800s emerged peat extraction and processing industries, in some cases the final nail in the coffin of the peatlands. But their history and culture mixed with older, local traditions. Ultimately industry and agriculture removed most peatlands; many were lost to drainage and farming 'improvement' from the eighteenth to the twentieth century. Peat fell out of domestic use as firewood and then coal, gas, oil or electricity became more convenient.

Peat bogs and fens are important in British culture, in the past valued but also feared, with bogs as black waters, places of doom and gloom. Neither land nor water, they are something in between and those unfamiliar with bog or fen found them fearful places where a misplaced footstep could mean slow death by drowning. Almost everyone has seen the villain sink into the Great Grimping Mire in Conan Doyle's *The Hound of the Baskervilles*, and many people still fear the northern moors as the place of the Moors

This Victorian print shows relief tickets being distributed to the poor in the turf market at Westport, County Mayo, in 1880. A loaded peat cart is visible, as are peats loaded onto panniers carried by a rather rangy looking horse.

A typical Irish rural scene called *Galway Gossips*, with peat-burning cottages and their ever-present smoke, and the attendant peat stacks, in Galway, Ireland.

An Irish peat market in 1880: a scene of hustle and bustle with peat barrows and baskets, and loaded peat carts.

Murders. These places are productive landscapes that supported people for centuries, but fearsome and loathsome to strangers.

One of the terrifying occurrences for those living near a major peat bog was the bog burst. This might be the result of the bog simply outgrowing itself to a point of instability and sometimes a result of climate. However, it was usually the consequence of human interference, frequently resulting from attempts to drain or cut the bog. This could destabilise all or part of the bog with dramatic and devastating consequences. From the sixteenth century to the 1900s there were numerous and spectacular bog bursts and slides, sweeping trees, houses, people and livestock along with them and sometimes extending over several miles.

Peat and bogs are imprinted on our minds and language with terms like 'bogged down' and 'mired down'. The idea of 'putting the dampers' on something, or 'dampening down', derived from placing a damp piece of turf on the peat fire last thing at night to keep the fire smouldering until dawn. Seminal moments in history are linked to bogs, fens and peat. It was in the bogs and fens of Sedgemoor and Athelney that King Alfred held out against the Viking invaders; and the perhaps apocryphal cakes he burnt on the fire would have been oatcakes on a peat fire rather than the singed coffee cake of popular imagination. But this is an image emblazoned in the mind of every school child taught English history, and at its core is a peat fire smouldering away. Hereward the Wake, the last great Saxon leader, maintained his stronghold against Norman overlords in the Isle of Ely, a small area of dry land in a sea of fens, water and peat bog. And the last major battle on English soil – the battle of Sedgemoor in Somerset – was fought, won and lost in the middle of a great fen.

In Ireland there are early references in the few written sources to early Irish life. In particular there are extracts from the *Senchas Már*, the Old Irish

The Peat Market or Turf Market, Galway, Ireland. Peat traders and housewives are buying peats.

The 'coal of the country' with dried cut peat being stacked near the peat-cutting grounds in Ireland. Once fully dry and when required, it will be carted away to the market or to individual domestic stores.

law text. There were also significant fines for illegal peat cutting and in cases of accidental drowning 'the ditch of a turf cutting' was excluded from liability.

Most turf was cut as slices in a relatively dry and structured state. However, if the peat was too wet and problems of drainage were too intractable to dry the working area, it was extracted as a sort of mud. In the bogs of Cavan, Leitrim, Down and Sligo this was described by Boate in the eighteenth century: 'On that dry place where the mud is poured forth, sit certain women upon their knees, who mold the turf, using nothing else to it but their hands.' This 'mud turf' was still being made from peat taken from Ardee Bog in County Louth in the 1980s; it was then taken down the River Dee to be sold as fuel in Ardee town.

The family chief of the Clan Rose of Kilravock settled in the county of Nairn at the time of King David I of Scotland. The clan's main residence has been the Castle of Kilravock since 1460, and their clan symbol is the turf cutter.

WHAT IS PEAT AND HOW IS IT FORMED?

Peat is undecomposed vegetation formed in conditions of waterlogging or high rainfall. The moisture prevents air getting into dead plant material and over time part-decomposed 'peat' develops. The type of peat depends on the environment in which it formed and the plants from which it is made. Vegetation changes over time through natural progressions called 'ecological successions' and so the type of peat formed in any particular location changes. So a cut peat at one location, a 'peat profile', varies with depth, the lower profile being more fully decomposed and densely compressed. Peat formed in a fen or carr (wet woodland) environment is from reeds and sedges, relatively rich in mineral bases such as calcium, and slightly alkaline; that from sphagnum ('bog mosses'), low in mineral bases, is acidic. Lowland fen peats are the former and upland blanket bogs or mires, the latter.

The anaerobic, often acidic conditions help preserve anything organic falling onto or into a peat bog. This explains 'bog bodies' in peat bogs and the less widely known but scientifically interesting deposits of plant pollen and remains of insects from centuries or millennia past. These deposits are one of the main tools we have for 'reconstructing' images of past landscapes and ecologies to help to understand long-term environmental changes. When peat bogs are destroyed this unique information is lost.

Peat is influenced by the wider landscape in which it is formed, especially the rocks, groundwater and surface waters. Mineral-rich waters from limestone affect peat composition, and coastal zones have a saline influence.

In some situations, now rare, peat-forming successions reach their ultimate pinnacle with great 'raised mires'. Formed after a sequence of peat development on expansive flat landscapes, undecomposed peat builds to such a point that it rises above the surrounding water table, so the only water input to the upper zones is pure rainwater. These systems are incredibly low in nutrients and develop a uniquely interesting but very vulnerable ecology. To walk across such as site is an amazing experience; neither land nor water, the surface is the consistency of a huge jelly, the whole wobbly mass quivering and trees shaking. Most sites have gone, and those that remain are reduced to a fraction of their former extent.

WHAT ARE THE PEATLANDS?

By medieval times peatlands and commons of various types were noted on surveyors' reports as 'wastes'. This does not mean they were unproductive: in fact far from it. These were wild, remote, but highly productive environments; sources such as the *Domesday Book* (1086) show that lowland fens in particular created considerable wealth. These were areas beyond farming cultivation with available technologies, and they were difficult to control politically. They were productive for local people but not for the crown or big landowners. Peatlands provided fuel, food and building materials for human populations, and grazing for domestic stock over centuries, across much of the rural environment. The term 'waste' reflects perceptions of wealthier land-owning classes that these areas were not productive for them directly and by the nature of their terrain were dangerous places. They harboured unrest and disease and, for those living outside, the great fens or heaths were strange and fearful places. Extensive

A young Irish lady with peat barrow and the family pig in the background.

The Norfolk Broads, an abandoned medieval peat cut, are now a major tourist attraction. The broads were to be a naturally formed wetland area until research in the 1950s showed they were generated both through massive peat cutting for fuel (largely for Norwich) and by the effects of a rise in sea levels.

lowland and coastal wetlands merged with riverine marshes and swamps, with rolling moor and heath from the lowlands up to higher grounds. All these had large areas of peat. For the ordinary people, the commoners and peasants, and for some landowners and later industrialists, this was an important resource.

Given favourable conditions, peat will grow on a site, but always slowly; exploitation may mean the gradual eating away and reduction of the core resource. To gain access to very wet areas requires either drainage or dredging (as was employed extensively by Dutch peat 'miners' in medieval times). Drainage results in drying and peat shrinkage and loss, and facilitates agricultural encroachment. By the nineteenth century, the demands for new and 'improved' agricultural land meant peatlands came under direct assault, and in two to three hundred years were almost totally obliterated. Those remaining are fragile and vulnerable environments, ranging from productive, lowland fens to low-nutrient, unproductive peat bogs, moorlands and heaths. Subsistence use in remote areas of the Scottish Highlands and Islands was often not sustainable as local demand outstripped supply or the ability of the bog to regenerate itself. Human suffering and depopulation resulted. Similar trends occurred in Cornwall and Dorset in England. In the depths of the medieval 'Little Ice Age' these areas suffered extreme abject poverty.

WHERE IS PEAT FOUND?

In the British landscape before widespread enclosures and 'improvements', peatlands occurred almost everywhere. By early 1900, this had changed dramatically, leaving a tattered rump of this once expansive resource. Most extensive peatlands are now found in upland areas of moorland and blanket mire, from Dartmoor, Exmoor and Bodmin in the south-west, to the Pennines along England's spine, and then the hills and lower mountains of Scotland. Along the western seaboard, especially of the Scottish Highlands, in Ireland itself and on the islands around both there are extensive blanket mires and raised bogs. Tracing the great river valleys and in particular in their lower floodplains (such as the Solway Firth and the Flanders Moss complex around the River Forth in central Scotland) are also remains of once much greater peat bogs. In lowland and more southerly areas such as the Cambridgeshire Fens, the Cheshire and Shropshire Meres, and the Somerset Levels there are relatively extensive peat bogs. On drier heaths and moors are shallow peats and organic soils. Detailed historical research has shown former peat bogs and fens in areas where today it would be almost unthinkable.

PEAT OR TURF?

Confusingly there is no satisfactory definition of these two terms in relation to the fuels and materials to which they refer. The site of peat extraction can be termed a turbary, a turf pit, a peat moor, a fuel allotment, or very often a peat moss. In many areas the place-name of the local settlement or manor is attached to the local peat moss and this is the turbary for that area: for example in Lancashire there are Leighton and Leighton Moss, Silverdale and Silverdale Moss. The material itself is sometimes separated into 'peat' for the deeper stuff, and 'turf' for the shallower tops, but this is not always so – sometimes 'turf' is also used for the deeper peats. There are also localised and technical names often with subtle meanings in terms of brown fibrous peat from the higher levels and deeper, hard black peats from lower down. Turf is always the name used for thin heather and grass sward cut from drier sites and burnt, and the same implement, a turfing spade, was used for horticultural grass turf as well. These thin turves were generally considered a poor fuel, but they did burn hot with high silica content from sandy soils, and in some areas were quite prized.

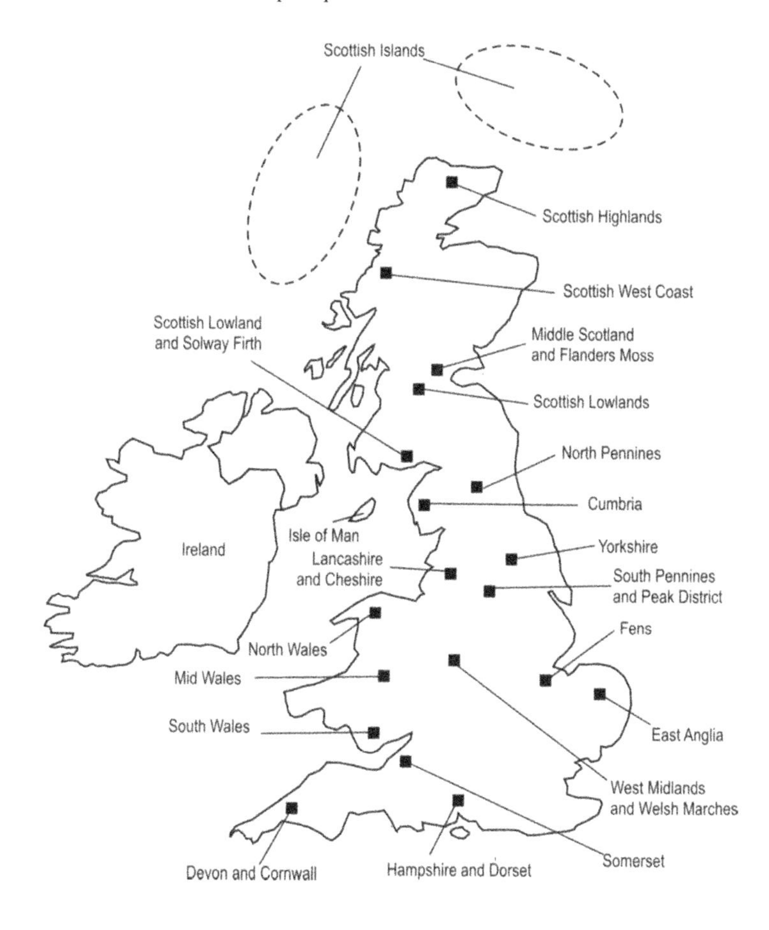

Map showing the main areas of peat and turf use in the British Isles.

USES OF PEAT AND PEATLANDS

THE POTENTIAL to exploit the use of peat, bogs and fens taxed the minds of scientists, industrialists and farmers for generations. Some were enthusiastic and others less so. According to Robert Alloway in 1869, 'few are now disposed to believe in the practicality of the profitable use of peat but on the contrary are more inclined to consider with the *London Times* that the one huge succession of bogs which fill up a large portion of the central parts of Ireland, are useless and unsightly excrescences, growing over and covering up three million acres of good land.' However, for many people these were productive landscapes, and before their almost total destruction through enclosure, 'improvement' or abandonment, most people – especially the poor – relied on them for a diversity of products.

Fuel: peat, turf, ling (heather), gorse or furze, kindling, birch coppice or brushwood

Building materials: peat, turf, ling, stone, gravel and sand, sphagnum moss, bracken or fern, rush, clay, birch poles, other wood

Grazing: sheep, cattle, ponies or horses, deer, rabbits

Other foods and herbs were gathered.

Most people were labourers, farm servants, squatters and cottagers; the last owned or occupied cottages, with ancient customary rights of commonage on the wastes. These included the right to pasture one or more animals on the common, to cut turf and to extract fuel. Massive conversion of heath, moor, waste and marsh to arable and enclosed pasture dramatically ended this utilisation of the landscape by the majority of the rural population: the rural poor and the poorer commoners. Along with wood mostly from 'woodland' and wooded common, peat, turf, gorse or furze, broom, bracken, fern or brake, flag, reed, sedge and ling were harvested from heath, common and fen. If it occurred at or close to the surface then mineral coal might also be taken. It was generally not favoured because most coals generated noxious fumes when burnt in a primitive way. Fuels might be combined as need and availability dictated. Gorse or furze was often used as an alternative if wood and peat were in short supply, or as kindling or for high temperatures such as baking.

Opposite:
A team of workers cutting turf on the hills of Donegal. Working the cut face of a blanket bog, the men are working in pairs with one cutting and one throwing.

Crofts and peat stacks in Weisdale, Shetland. Note the open bleak landscape and the lack of trees for fuelwood or shelter.

The one-time extent, the cultural history, and deep-rooted significance of these landscapes are generally now forgotten. Much of this landscape was used for peat cutting, 'turf-getting', 'moss-gathering', bracken and rush harvesting, and other uses. For upland landscapes especially, this was far greater than previously suspected. For example, there are extensive well-defined peat cuttings around remaining areas of blanket mire in the Peak District and similar areas such as the North York Moors. Thin peats and turf on low-level moors were stripped away with extensive 'turf-cutting' and 'paring and burning' in upland landscapes and on valley-sides and down to the lowlands. Loss of the resource was for many local communities a devastating blow.

Delivering peats to the customers in a Somerset town. The lady is carrying 'Dree Penn'oth' or three pennies worth!

Small-scale use lingered on in remote areas, particularly the uplands, but by the late 1900s almost all subsistence exploitation in England and Wales had died out. Traditional use in Scotland and Ireland is also in rapid decline.

Even when landscapes have changed and use stopped, a legacy of place-names remains, for example Peat Pits, Peat Lane, Peat Road, Turf Field, Moss Crop, Moss Side, Turvin Moor, Turf Moor and Turf Hill. Surnames – Peatman, Peat, Moss, Mossman, Peatfield and Peate – also remind us of our peat-cutting history.

PEAT AS FUEL FOR THE COMMON PEOPLE

This was the most widespread use of peat, occurring almost everywhere. Peat and turf were not the only fuels from heath and common and use of materials varied from site to site with ownership and rights, and particular needs. John Evelyn, writing about 'fuel', noted in 1729:

> 'But besides the Dung of Beasts, and the Peat and Turf which we may find in our ouzy Lands and heathy Commons for their Chimneys, Cow-sheds, etc. they make use of Stoves, both portable and standing… In many Places (where Fuel is Scarce) poor people spread Fern and Straw inn the Ways and Paths where Cattle dung and tread, and then clap it against a Wall till it be dry…'

For many commoners and other peasants, peat and turf were the main fuels. Around Milnthorpe in south Cumbria for example, peat was the main fuel until about 1800. An article of agreement made on 11 May 1687 between Mary Hinde of Auckenthwait, Spinster, and Thomas Cragge of Cragyea in Auckenthwait concerned 'a dwelling house and garding', and stated 'that Thomas should buy the property for £8 providing that Mary Hinde shall use occupie and possess ye Chimney and ye body of ye aforesaid house during her naturall lyfe… and that ye said Thomas Cragge shall give and lead cart to ye doore of ye said Mary Hindes house five cart full of peats or turfs in every yeare during her natural lyfe.' Free use of the fuel was important to survival.

FUEL NEEDS

We need fuel to live and as our settlements grew from villages and farmsteads to towns and cities these demands have grown and become more complicated. How to keep warm and to have fuel to cook have always been at the core of human survival. Only in recent times have we had the ease of electricity down a cable, gas or oil down a pipe, or even coal delivered to the door. In times past, as recently as the mid-twentieth century for many people, you went and got your fuel from the local peat bog, your turbary.

Today we talk of 'sustainability', a delicate balance between people and their environment. In history where this failed the consequences were devastating. The Western Isles provide examples of fuel and soil exhaustion and consequent starvation. In England, the urban population could resort to alternatives such as coal but rural areas could not unless supplies were close, and until the 1500s most mineral coal came from coastal outcrops and coastal transport. So the rural population turned to other fuels as they always had, with peat, turf and furze, supplemented by small wood and heather, the main fuels of common people. Fuel was important and often scarce, and of course from around 1500 to the nineteenth century the climate in Britain was incredibly cold. Life was hard and without fuel people died. The turf consumed by a single household for fuel, litter and other purposes was around 8,000 turves per year; 6 tons of peat was equivalent to 1 ton of coal, with peat stacks as big as the cottages. At Martham in East Anglia, a household used 5,000–8,000 turves per year, and at Scratby 10,000 per year.

Alongside domestic use peat was taken commercially, the most dramatic example being the Norfolk Broads. Long considered 'natural', in the 1950s they were identified as abandoned peat workings. They supplied Norwich and its cathedral priory with 400,000 turves per year during the 1300s and 1400s. Similar use occurred in the Peak District and South Pennines, Paul Ardron showing that more peat was cut from the South Pennines (about 34 million cubic metres) than the Norfolk Broads at the same time. Most was in the early medieval period, with lower-lying sites taken during the sixteenth, seventeenth and eighteenth centuries, associated with Parliamentary and private 'enclosures' of heath, moor, common, bog and 'waste'. According to Young (1804) the fuel allotment was 1.75 acres at Northwold in Norfolk and was supposed to produce 12,000 turves a year, the calculated consumption of one hearth. Cottagers of twenty years' standing with no common rights were treated meanly, being awarded permission to cut only 800 turves each year, controlled by the fen reeves (local administrators of common rights), quite insufficient to keep a hearth alight. According to Wright (1964), when wood was scarce in the UK:

> Peat was the only alternative fuel to be had in any quantity. There is even more of it in Britain than in Ireland, despite the poets and the travel brochures; but it was too bulky to be carried far from its source. Peat burns readily; its merit is to smoulder without a blaze, though this makes for a smoky fire. The 'peat-reek' is pleasant at a distance, and a whiff of it is not unwelcome in one's Harris tweeds; but in a 'lum' cottage its pungency is dreadful. Peat has been 'coked' to destroy the reek, powdered, mixed with pitch or rosin, and compressed into bricks that were claimed to be better than coal. Lacking true peat, some burn turf, or parings of peaty soil with roots of heather and gorse. (Peat and turf, accidentally ignited, have caused

slow, widespread and unquenchable fires, which have even consumed whole villages.) Dried cowdung is a good fuel, and the scent of its fire sweeter than might be supposed; it found some favour during the wood famine, but to burn dung that ought to enrich the soil is bad husbandry.

The Reverend William Harrison, writing in Tudor times, was concerned about the depletion of woods and the lack of fuel:

> Howbeit, thus much I dare affirm, that if woods go so fast to decay in the next hundred years of Grace as they have done and are like to do in this … it is to be feared that the fenny bote, broom, turf, gall, heath, furze, brakes, whins, ling, dies, hassocks, flags, straw, sedge, reed, rush, and also seascale, will be good merchandise even in the city of London, whereunto some of them even now have gotten ready passage… Of coal-mines we have such plenty in the north and western parts of our island as may suffice for all the realm of England; and so they do hereafter indeed, if wood be not better cherished than it is at this present … their greatest trade beginneth now to grow from the forge into the kitchen and hall, as may appear already in most cities and towns that lie about the coast, where they have but little other fuel except it be turf and hassock.

Peat and turf were often 'free' for the cost of cutting and transportation, and a valuable resource to help support vital and potentially expensive services such as the schoolmaster in Scotland. The schoolmaster '… also has his peats cut, dried, and brought home free', according to the Reverend Mr

The traditions and skills of working the peats were passed by word of mouth from generation to generation. Here young peat workers in Ireland are getting their first experience of harvesting and processing turf.

James Dingwall (*The Statistical Account of Scotland 1791–1799*, Parish of Far [County of Sutherland]). Suffling (1885) described the value of peat:

> ... This is turf, and its lower brother peat. This peat, or, as it is here called, 'hovers', is, when properly dried, a capital and economical substitute for coal. It gives off a blue smoke when burning, and this, as it rises from the cottars' chimneys, wafts a rather pleasant perfume in the air, which is a great improvement on the soot-laden, evil-smelling smoke of the metropolis. A peat-ground, properly managed, is a rather valuable holding, as may be gathered from the following statistics. The peat blocks, when cut are about 4 in square (shrinking by drying to about 3¼ in) by from 2 ft to 2½ ft long (the depth of the boggy surface soil). Each square foot, therefore, produces 9 'hovers', each yard 81, each rod 2450; and, consequently, each acre the enormous number of 392,000 hovers. As these are retailed at from 1s. to 1s. 6d per 100, a good profit must be realised. [Value = £196–£294 per acre.]

Thorold Rogers noted the use of turf in 1337, with amounts as the 'thousand' (in fact 1,200) or the 'last' (12,000). In Cambridge in 1334 the turf was specified as heather, average price around 9d per thousand. Turf was used frequently in Southampton and in Kent. Between 1415 and 1450 the Cambridge colleges used turves extensively, bought by the thousand at around 2 shillings.

FUELLING TOWNS, CITIES AND INDUSTRY

Peat was the fuel of the rural poor, but used in towns and cities too. This often finished so long ago that few records or memories remain. However, along with the obvious example of Norwich, peat and turf were used in Carlisle from the Solway Firth mosses, and in Cambridge from the Fens.

This picture of turf cutting in Ireland shows the peat face, the peats raised up to dry, and the carts taking peats away.

In this image we see the carting of peats from a riverside peat moss at Milnthorpe, Cumbria.

York was fuelled from Askham Bog and other nearby turbaries. There are interesting records of peat turf used in York. In 1388, peat or 'turbarum' was being brought into the city by water to supplement York's own turbary on the Tillmire near Heslington. By the seventeenth century the amount of peat turf being consumed increased with improved transport along the rivers and small boats carrying peat from Thorne Moors in South Yorkshire.

In York in 1643 there was reference to the selling of turf, when nine men were accused of 'selling turfs contrary to my Lord Mayor's price'. Admitting guilt, they were fined ¾ penny each '... according to an Order made in the like case the 9th day of November 1593'. This latter reference was to four men reprimanded and fined for a similar offence. In the 1643 entry for fines paid to the Corporation of York it is stated that money was received as '... of watermen for selling turves before the price was sett by my Lord Mayor 8/-'. The notes refer to the month of November, so perhaps a time when the cold was making turf especially important. Peat was worked across a wide area of the old West Riding during the seventeenth century, with references indicating 'Turf Mosses', to 'wayne leades of dryed peates', to the use of carts for carrying dried peat and to 'the turfpitt' from places such as Giggleswick, Thruscross, Barwick-in-Elmet, Bentham and Fishlake. In the latter case the notes refer to the theft in 1652 of 'one Catch loade of Turves & wood' with a total value of £7. This was by four men, one of whom was referred to as an 'Airmyn Waterman', and the 'Catch' was a ketch, one of the traditional open sailing trading barges that plied up and down these watercourses. There were even turbaries near to present-day Leeds.

Peat and turf fuelled industries, mostly metal smelting. Suitability varied with the task – sometimes it was used as back-up fuel when wood or charcoal was unavailable.

HOW PEAT WAS WORKED

THE FIRST STAGE in peat production is drainage to make the site workable and safe. The next step is to develop access to get people and equipment and often animals (such as donkeys or ponies to carry the peat) onto the peat ground or turbary. There need to be working areas and places to rest and eat. Working areas needed spaces for treating cut turves or peats, and stacking them to dry, turning them whilst drying, and then drying in larger stacks. Finally, peats are carried off site by sledge, cart or pannier and stored in covered stacks or barns close by the cottage.

DRAINAGE

To extract peat from a bog or fen, the site must be relatively dry. There are some exceptions to this, for example the Dutch using dredges to take wet peat from the polders, but in Britain this was localised. Instead, water tables were lowered permanently or temporarily. This is generally easier on high ground where the key is to cut a drain and take water off downhill. Individual bogs can then be drained at least in part and dry edges worked or cut for peat. In low-lying areas this becomes more difficult: unless you can drain an entire landscape, your neighbours' water flows back onto your land. To undertake effective drainage in the lowlands requires control over large areas of land, through ownership or community co-operation. You need the necessary technology and skills to cut drains and to move water from the flat land into these. Drains are often higher than some areas of the surrounding landscape, and so it is necessary to pump water off the land into the drain. The small drains must then flow seaward and into larger drains or canals. Hopefully, the network of drains and dykes across the land will carry excess water away.

Once the land is drier, then it is possible to cut and extract peat or turf from the fen or bog. This is then carefully dried and eventually carried away for storage or use. However, serious difficulties can arise, which is why many areas remained unreclaimed and unworked for centuries. One obvious problem, which apparently did not strike the early drainers, was that if you

Opposite:
The photograph shows the neatly organised approach in a well-managed peat bank at Burwell in Cambridgeshire. The man is cutting peat with a Fenland beckett – the shape of the tool is very like a cricket bat.

Above: The conditions in Holland meant that peat was often dredged rather than cut, as here in the Dutch polders.

drain a bog or fen it dries and shrinks. This means that very quickly your site is lower than the surrounding land and water flows back in. The only solution is active pumping to a drain that is now definitely higher than your original ground. To do this requires pumps, originally wind- or animal-powered, then steam, petrol or diesel, and then electric. The water is then carried to great channels with high embankments and off to the sea.

The final problem is that in landscapes like the Fens or the Somerset Levels, there is little 'fall' from inland to the coast. This presents problems in getting water to flow, and leaves you vulnerable to inundation from landward or seaward in the event of a storm, or worse a rise in sea level. Time and again this has been the fate of such 'reclaimed' lands. The Norfolk Broads, a massive medieval peat cutting, was finally abandoned to water as the land shrank and sank, and sea level rose.

Right: This picture shows the loading of a barrow with peats cut at Burwell in Cambridgeshire. Note the open sides to the barrow and the high front to hold a good load, and the simple but broad wheel to cope with the soft peat surface.

Aerial view of Burwell peat pits in Cambridgeshire. Field boundaries of the enclosed and drained fen are clear, as are the ramshackle buildings to house equipment and perhaps to dry cut peat.

In regions like the Somerset Levels and the Fens, the modern landscape was imposed on the extensive wildlands of fen, bog and water between the eighteenth century and the twentieth century. Much land was 'enclosed' by Parliamentary enclosure with each administrative area (generally a parish), requiring its own Act of Parliament. Within this mechanism there was the possibility to establish areas such as the fuel allotments and poor lands. Those with common rights often received small pockets of enclosed land as part compensation for the loss of their privileges. In some areas this led to a complex system of drains or ditches (in Somerset called 'rhynes') that divided the land into small rectangular units and occasional odd-shaped leftovers. This gave the characteristic landscape patterns with roads and farms (and smaller plots too) all following this basic layout of large individual landowners. Some of the plots were essentially peat bog or fen and still very wet with several metres of peat. By the late nineteenth century, many sites were being worked for domestic fuel but also as commercial businesses. Much was for peat fuel, but there was increasing demand for animal litter and growing markets for horticultural peat. During the mid to late twentieth century there were moves from traditional hand cutting to commercial machine milling and destructive opencasting, with traditions lost almost overnight.

TOOLS

Between early enclosures and the industrial late twentieth century, there was a period of sophisticated and complex hand cutting. The techniques and implements used evolved over centuries, probably from before Roman times, although it was they who invented the primary tool of this trade – the spade. There were wooden and perhaps bone peat spades used in earlier times in both the Cambridgeshire Fens and the Somerset Levels, but a spade made of iron or steel was the invention of Roman military technology. Given the basic tool, over the intervening 1,500 to 2,000 years a diversity of regional approaches and implements evolved with distinctive local character. The techniques and tools were similar from lowland Somerset, to Cumbrian valleys, Irish bogs or Scottish isles.

In the Fens the main tool used was the 'beckett', a slim straight spade like a cricket bat but with a flange at right angles to the main blade. This made two cuts in one movement and was pushed easily through the peat by hand. The many differing shapes and sizes were locally and individually made with willow shafts and blades from cast-off iron tools and other scrap. The blade had wood exposed to help turf slip off more easily, and the flange was around an inch less than the blade width. The result was a sharp-bladed tool that cut neat brick-shaped turves that were easy to handle, carry and dry. Dried turves were sold by measure, in the Fens based on a small beckett used at Isleham Fen in the 1850s. This measured 14 inches by 3 inches. Small turves dried quickly but took longer to cut. At Wicken a blade was developed that was around 18 inches by 4½ inches, producing (after shrinkage) a turf 11 inches by 3½ inches. Isleham small turves were sold in a thousand and Wicken large turves were sold as six hundred 'to a thousand'. The Wicken beckett size and measure were subsequently adopted by turf cutters across the Fens.

Below: The cutter is working the turf bank at Burwell in Cambridgeshire with a typical Fenland beckett.

The earlier Fenland tool was the 'moor spade' or 'sharp shovel' designed for use with a strong push from a foot. Some were made totally from iron, making digging hard and slow. The digger made out the line of his cut and then used the spade to clear vegetation from the top 15 inches or so. The next step was to 'crumb up' or shave the exposed peat to produce a level cutting platform. Vegetation around the side might also be cut away to facilitate stacking of the cut turves. A 'peat knife' or 'turf knife', basically a crude metal cleaver, might be used to cut down through the sides of the pit or trench. Either this, or the cutter would begin with the beckett immediately. He made two cuts for each turf and worked backwards, cutting one spit down (that is, the depth of the beckett) and four spits across. Each turf broken off was lifted easily to the side of the pit. The trench would be cut the length of the field being worked, with the surface cleared and thrown back into the previous pit. He then worked back to widen the pit to four spits; with turves opened and set back, dug along the other side. Depending on the field and the number of men working, he might open another pit, but with room between trenches to allow barrowing and stacking. Once dug, the individual field was abandoned for perhaps thirty years before being dug again. The site would have levelled off and new peat formed. However, it was not new peat that the cutter was after but the deeper old peat, perhaps a further spit depth below the original cut. Some sites produced two or three spits, and exceptionally four or even five spits depth. However, pumping water out by windmills, and holding floodwater back by dams, became real tasks.

Individual turves weighed about 7 pounds, and the pit was 1 yard wide. The digger lifted to his side to leave three turves in height and broken ones on top. These were commercial piecemeal diggers, so speed, quality and consistency were vital. The operation was more difficult once diggers moved to a second spit depth and a narrower trench (only three spits wide). Around 1900, the piece-rate was 5 shillings per thousand turves, including opening, digging, turning, barrowing, stacking, and boating. Digging and laying were paid at 2s. 6d per thousand (that is, a Wicken 600). Men working from 6 a.m. to 3 p.m. produced about 3,000 Wicken turves per day. The highest recorded digging rate was an amazing 6,000 per day, but work was often restricted by poor weather.

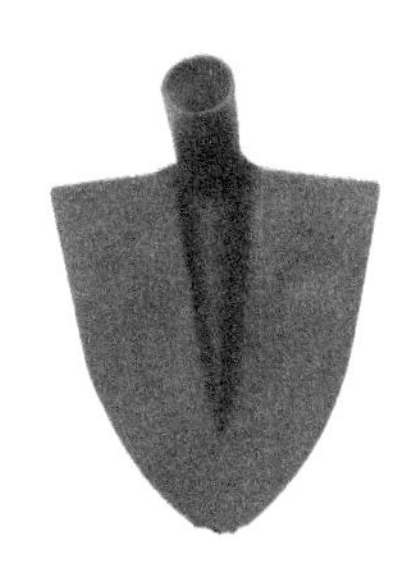

Top: A typical Devon spade as used across the whole of the western peninsula.

Above: This picture shows the blade of a typical Fenland peat spade rather than the beckett used for cutting.

PROCESSING AND DELIVERING PEAT

Once dug and stacked, turves were left to dry for around three weeks, depending on the weather. The turves were opened or dressed, which meant re-stacking with gaps to allow air circulation to speed drying. Each turf had to be handled individually, but a skilled digger could dress around 3,600 an hour. After perhaps six weeks the turves were sufficiently dry to barrow and stack by the drain. Since the better drying weather was late summer, the priority for early work was digging. Final shaping of turves was with the beckett or a special trimmer. Those not up to standard were cast aside as 'bits' to be sold cheaply to local people.

This was physically demanding and thirst-generating work in an airless peat pit with summer heat, high humidity and thousands of biting flies. The water of the bog and surface drains was unsuitable for drinking but diggers slaked their thirst from the clear water of drainage 'lodes', ditches and canals that facilitated drainage and transport. With bad weather and in deeper pits, diggers in greased leather thigh-boots might stand in water for part of the time. To avoid slipping they wore broader boots over the thigh-boots or short wooden boards strapped beneath the soles. They used flat-bottomed stilts strapped to their calves.

Once stacked, turves could be soaked again by heavy rain and need further turning and drying, often by young lads. They also did a lot of the barrowing and learnt to cross drains and bog along wooden planks and trestles. Their open wheelbarrows measured 6 feet from the front of the wheel to the end of the handles, and carried sixty Wicken turves a load. The turves were stacked on a bed of litter to keep out worms and other problems. Marked with a reed or a missing turf every thousand, turves were taken straight to the boats or stacked and stored for later shipping. The planks along which the barrows were trundled were around 14 inches wide of narrower spars bolted together, and slippery, flexible and bouncy; the lads wore boots with spiked clips to help them grip. The men barrowed around two hundred turves at a time – the youths less – and each small boat carried around 2,200 Wicken turves. The next task, with due heed to water depth, was ferrying the flat-bottomed boats of peat pulled by trained donkeys along the lodes. Sometimes stacks of peat were left near the turbaries to be ferried away next year. The boats carried peats to storage sheds owned by major operators paying annual mooring fees to the parish council. It was stored in sheds when space allowed, with turves placed on wooden slats to allow air to circulate, or left exposed.

With turf needed all year round, the supply to households was an important business. At Wicken, turf hawkers came from Soham, Fordham and Burwell to fill their carts. They took 1,500 turves at a time in small carts pulled by ponies or donkeys. Around the immediate neighbourhood of a turf-digging village such as Wicken the diggers themselves would sell and deliver peat. Some locals got their peats direct from the storage sheds, buying

These girls are busy loading peats and perhaps gathering and turning them to dry at the drying grounds. Note the other women working in the distance in this extensive turf-cutting area.

quarters in a sack or more in a barrow. Diggers' wives helped unload turves from the boats and loaded hawkers' carts. In later years peat might be bought off the back of a lorry.

Blade of a heavy peat spade and draining spade.

LOCAL VARIATION

The approaches and techniques applied to turf cutting throughout the British Isles were a mix of those applied in lowland fens, or bogs and uplands of the north and west. There were many local variants and distinctive local approaches, too variable to describe, and over time they have been mingled and altered. In recent decades tools and implements have moved widely as people have travelled.

ENGLAND

In Somerset, the peat cutters cut a 4-foot-wide 'head' along a line marked out with pegs. The 'unridding' (top turf and rubbish) was thrown in the pit, which was around 12–18 inches wide. The Somerset cutters then dug around four layers (about 3 feet 6 inches) deep, the peats called 'mumps'. They used turf scythes to mark out and dig, but in later years hay knives and spades, and then a mix of shovels, mowing scythes, marking sticks, choppers and others. As in the Fens, water coming back into workings was a problem ultimately solved by pumps. In the early days the diggers physically ladled the water out, 'lecking out' with a long-handled wooden scoop or shovel. Similar tools were used in early medieval times for shovelling out peat from underwater. A typical spade-like hand tool was imported by Dutch workers into Fenn's and Whixall Moss on

Peat cutting on Brown Willy, Bodmin Moor, Cornwall, with two men and a horse-drawn sled operating in saturated conditions. The cut turves are probably being taken off the peat moor to dry elsewhere.

Top right: Turf cutting in Somerset, showing the peat bank and working area. This picture indicates the intensive and highly organised nature of the extraction.

the mid-Wales border during the 1920s, and a similar implement called the 'Devon spade' or 'Cornish long-handled shovel' was used in south-west England. A shorter 'hay spade' or 'nicker-out' was employed to mark out the side of the peat bank for the next cut, to mark out the benches and to top the 'Dutch blocks', as the turves were called at Fenn's Moss. They also used a broad straight-edged tool called a 'sticker' for marking out. A widely used tool for taking off the top layer or for skimming thinner peats was a long-handled turf spade, still manufactured today and known as a 'paring iron'. In the Isle of Man these were originally called 'scraa-cutting spades', and were used to cut thin turves as tiles or thatch. Today they are still used to cut grass turf. After cutting the turves, there were two further processes in regions such as Somerset, called 'hiling' and 'ruckling'. The first was laying turf on the ground and building heaps of twelve turves called 'hiles' (or 'walls' on Crowle Moors on the Lincolnshire–Yorkshire border). The second was dry turves stacked as 'ruckles' or 'cocks' in Somerset, or 'pyramids' at Crowle, and 'peat stacks' or 'ricks' elsewhere.

Another spade was used for draining with variation, from square, rather narrow blades to broadly sweeping curves with an extended cutting edge. These were also used for peat cutting and site preparation.

SCOTLAND AND IRELAND

In Scotland and Ireland, like in the lowland fens, peat cutting was an ancient tradition. It was associated in both countries with the periods of land 'improvement' in the eighteenth and nineteenth centuries, in the face of starvation, crisis in the rural economy, and problems of over-population. By the 1900s, the preparation for larger-scale cutting of peat bogs was undertaken by government bodies such as Land Commissions or Congested Districts Boards, or 'improving' landlords. They organised drainage and construction of access roads and tracks, and even the layout of the peat faces or 'facebanks' to be worked. A main drain was opened across the bog,

In the background of this picture of turfing in Somerset can be seen the loaded peat barrow, a peat spade and ladder. The small ruckles for drying are shown across the whole area.

deepened year by year until the bog was 'bottomed' by the drain. Sub-drains were cut parallel to the main drain and cross-connected to it. In working a raised bog the cutters described four different and distinctive zones or layers. The top of the profile was the 'clearing' or 'top scraw', below which was 'white turf' of undecomposed material, and then 'brown turf'. At the bottom was rich black fen peat.

The tools of the Irish peat cutters included a special implement called a 'flachter', 'skroghoge' or 'straw cutter'. This was used in the north of Ireland, imported from Scotland, to cut away the heathery 'scraw' (from the Irish *scraith*, a green sod) from the bog surface prior to cutting. These were very

This photograph of turfing in Somerset shows the larger, second-stage ruckles and a wide range of different spades and other implements. It also shows the working bank.

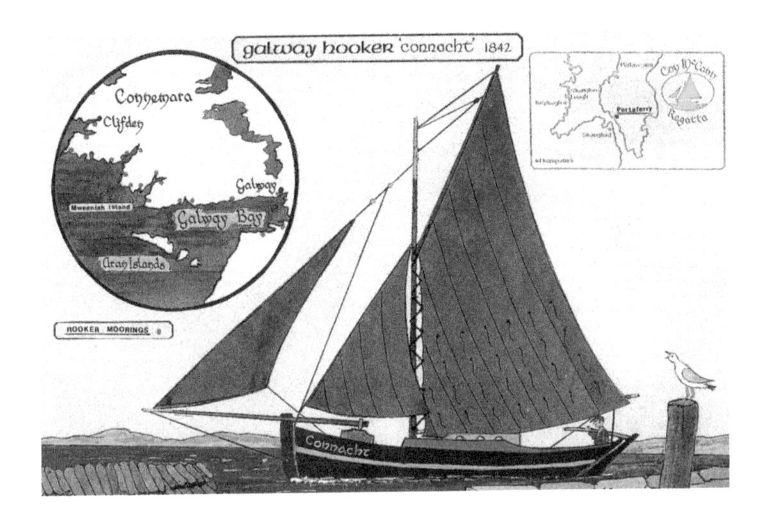

All around the western coasts peats might be carried from island to island or between islands and mainland. It all depended on demand and the potential for supply, which could become critical as local turbaries were worked out. The Galway hooker was used for ferrying peats around western Ireland.

individual tools made by the local blacksmith with a handle of pliable wood and a metal blade from scrap. The flachter was basically a modified breast plough and though not widely used in Ireland the process was sometimes important with up to 6 feet of bog surface needing to be cut away before turf cutting began. It was particularly used in the mountain bogs of north-east Ireland. In carefully managed sites the scraw was reserved and placed back on the bog after cutting. The Irish, like the Highlanders, have specialised vocabularies related to bogs and peat cutting, with regional variants. A turf bank in Scottish Gaelic is *bac móna*, *bakki* being Old Norse for bank. The Scottish Gaelic for turf spade is *tréigeir* from the Old Norse *torf-sceir* (turf-cutter). There are Norse influences through the Viking connections with Ireland directly and via the Western Isles. In Ireland most turf was cut with a turf spade or slane (*sleaghán*), similar to the Fenland beckett.

Scottish Islands peat spade with flange. The peat spade with flange is typical of those used in the north and north-western isles.

After clearing the top scraw the slanesman could begin cutting, the winged slane blade allowing a double cut with turf extracted in one swift action. Slanes were made locally by blacksmiths or in bulk in spade mills, with two main types. The 'breast slane' (winged or not) was used with the cutter facing the turf bank and required reasonably dry conditions as cutting moved along. It was used extensively in blanket bogs. The turves were cut with a horizontal stroke into the bank and then vertically by a second cutter standing on top of the peat bank. Cut turves were placed on top of the uncut peat with around sixteen 'boxes' of 14 × 6 × 6-inch turves per day. The 'foot slane' produced 10 × 5 × 5-inch turves and was used for deeper valley-bottom bogs. The sods were cut on two sides by vertical strokes of a winged slane, the bottom breaking as the blade lifted, and the sod thrown to the waiting barrow-man: quick and simple. The barrow-men took peats

An Irish turf cart is pulled by a donkey with two attendant teenagers in charge.

away to spread on the dry cutaway area or 'low bank', in front of the turf breast or 'high bank'.

In Scotland and the Islands, cutting was with a 'peat iron' or 'flauchter spade' made by a local blacksmith; these were similar to Fenland becketts or Irish slanes. According to Alexander Fenton this was noted in 1493, and the tool is even older. Peat cutting for fuel and to fertilise the Highland and Islands infields goes back further still. The tools are locally distinct, so in Shetland a slim straight-shafted spade with no foot-peg and operated by the arms was called a 'tusker'. Orkney and Caithness tools were similar to Shetland but sturdier and operated differently. A description from Caithness

At an unnamed site in Ulster turf is being loaded from the turf stack into a cart to be drawn by a small horse. In the distance the turf fields can be seen with cutover areas, the straight lines of the turbaries, and the associated stacks.

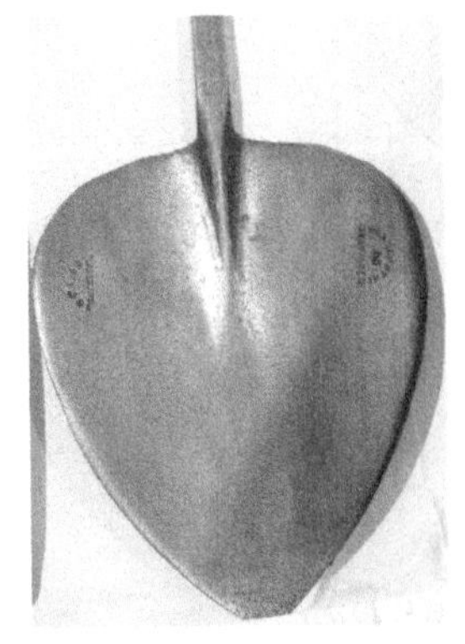

Above: In upland areas in particular it was common until relatively recently for farmers to use sleds for transport. Wheeled vehicles were expensive but also difficult to manoeuvre over rough terrain. The picture here shows peats being carried in a primitive horse-drawn sled in Ireland.

Above right: The blade of a flat paring or turfing spade.

in 1812 indicated two different ways of cutting peat, horizontal if peat was 1–2 feet thick, and perpendicular if deeper. The tusker or 'turskill' had a wooden peg to allow use of foot power. The cutter held the handle with both hands and pushed the foot to drive the spade into the moss. A woman stood at the base of the peat bank to take the peat as the cutter jerked it away from the bottom. She grasped the turf and threw it to the right side of the bank, where others, perhaps children, spread the peats to dry. Such good quality black peats were called 'tusker peats' and prized for sale; poorer quality turf was for home consumption. In Orkney, when the poorer quality hill peats were cut, a modified tool like a tusker was used. This had a broader blade with a vertical wing to one side, and was called a 'luggie', meaning the 'eared spade'. Across the Highlands and Islands is a great array of peat spades, some similar to basic tuskers, others with distinctive wings and flanges.

Left: The cutting and barrowing of turf, perhaps by a family, at Portrush, Northern Ireland. Cut peats are lying above the peat bank and the loaded barrow is pushed by a woman. Of the three men, one is working the top bank and two in the bottom.

In north-east Scotland, spades often had broad blades with a round mouth, and any wings were short. Two types of spade are described, the 'breist spade' or 'breasting spade' and the 'stamp spade' or 'underfoot spade'; the former was used horizontally, cutting into the peat bank face, the cutter standing below. The latter was handled vertically and stamped down with the foot from above the bank. Most cutters had both types for different situations. In central Scotland the pattern changed again: square-mouthed wooden blades were shod with iron cutting edges. Further into south-west and south-east Scotland there were local variants, though evidence for the latter is sparse as peat cutting died out long ago.

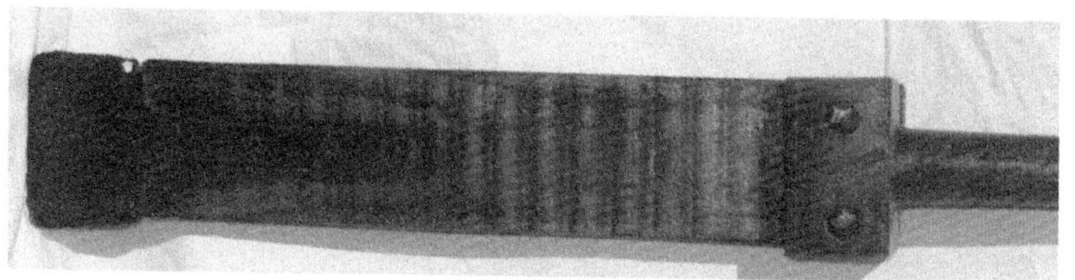

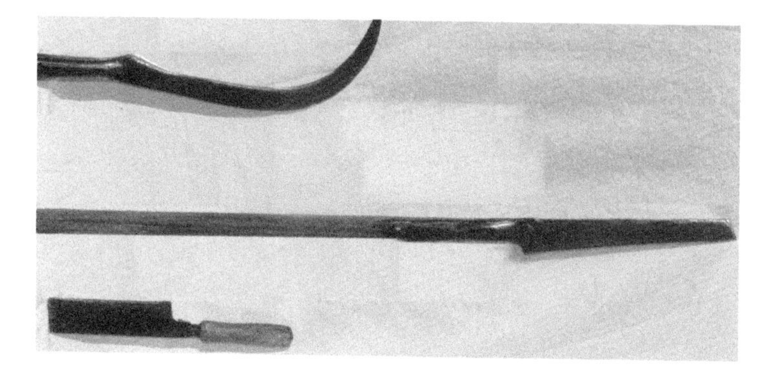

Far left: A turf knife such as this might be used to clean and tidy up cut peats or the surface of the peat face; a multi-functional tool, it would also be used to cut small wood and especially gorse or furze for fuel and fodder.

Above middle: The blade of a peat spade from central Scotland with its typical metal edge.

Above left: This typical peat spade from central Scotland shows the metal edge on the wooden blade.

Middle: Peat spades around the country show great variation in form. This is a Scottish Islands peat spade.

Bottom: Peat cutters used a wide variety of cutting, digging and lifting implements.

This picture demonstrates a very different set of tools (a long-handled spade and a turf barrow) and a different approach to cutting turf in Ireland from that in the English lowlands.

By the late nineteenth century, factory-made peat spades were manufactured in Scotland or imported from England; often reflecting local variants and demand.

PEAT IN THE ANNUAL CYCLE

In traditional communities the various tasks of living and farming followed an annual cycle based on seasons and weather, crops and activities, and availability of labour to undertake specific tasks. The cutting of peat and turf was a key part of the cycle carefully locked into other activities. Labour-intensive work happened when people were free to do it. In most cases the bulk of the work on site was from May to August, but varying with local climate, earlier on the Dorset heaths than in the Cairngorms. There are key stages of work: checking and preparing peat grounds was completed early on, so that when weather and labour permitted cutting could commence. Peat was cut, prepared and stacked to dry. Then, over the next few weeks, drying turves were carefully and regularly turned to dry thoroughly; this work, often done by women and children, was vital for next winter's fuel supply.

WHY PEAT CUTTING STOPPED

Turbaries were abandoned because they were exhausted or because cheaper, easier fuels arrived. In mid Wales, around Rhayader, the arrival of coal by train in the 1860s and 1870s meant abandonment of hilltop turbaries; valley-bottom bogs were exhausted long before. In the Yorkshire Dales and the North York Moors, extensive cutting probably finished by the 1930s to 1940s. Use was often not sustainable in the longer term and when peat ran out alternatives were found or communities moved on. The change was often through competition from other cheaper, more accessible or better quality

This peat bank in Shetland is being worked in a series of levels and the angle of cut into the top layer is clearly visible. Cut peats are laid on top of the bank and this is clearly blanket bog at a high elevation.

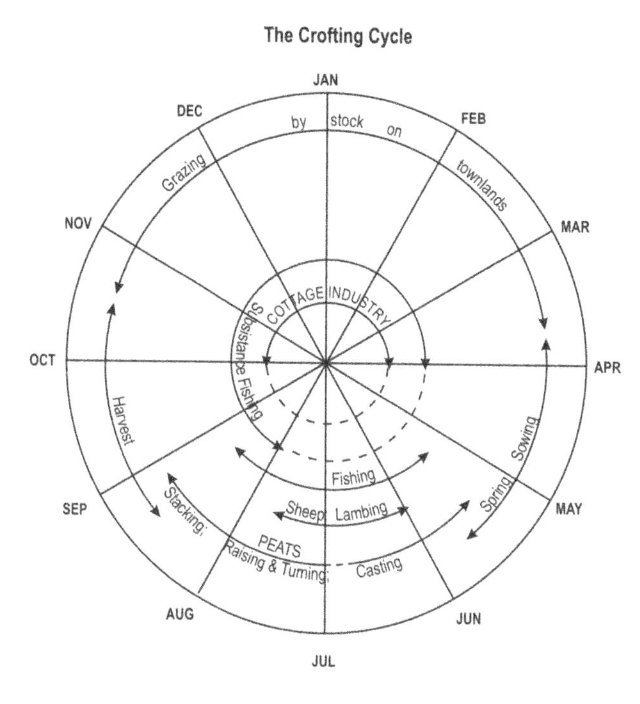

Diagram to show the cutting of peat in the annual cycle of work and an interpretation of the crofting cycle in northern Scotland. Based on Knox (1985).

fuels. Mann (1993) in *The First Statistical Account of Scotland for the Parishes of Anworth and Girthon, 1792*, notes that for coastal areas of south-west Scotland: 'Peats, the fuel used by the farmers and cottars, are dear, owing to the distance of the mosses, and, the bad roads which lead to them... Coals are the general fuel here. They are imported from Whitehaven, Newcastle, etc.

These young Irish ladies at the peat stack are taking a well-earned rest from the physically demanding work.

and run from 30s. to 40s. a ton.' Smith (1951) noted that in northeast Scotland, 'Peats are not used so widely as in former years. Prices are high normally £2 and over per cartload. Coal is brought to the door by motor lorry from Aberdeen and sells at the high price of £5 a ton for English coal and £4 to £15 for Scottish coal.'

Loss of demand from industries such as lead smelting in the Yorkshire Dales, and rural depopulation were other causes. Exhaustion of supplies, the hard work of extracting and preparing the turves, and financial considerations (especially cheap and easily accessed energy supplies from coal, electricity, oil and gas) were all important factors. Indeed, the survival of the practice to the mid twentieth century was often due to the fuel being cheap or even free. In some cases manufactured products from peat fuel were lower grade and lower value than from competing fuels.

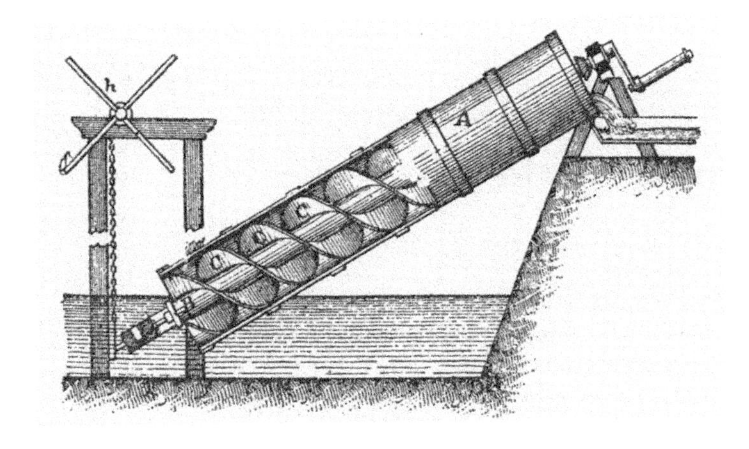

To cut rather than dredge peat, it was generally necessary to drain the site. In some cases this required sophisticated pumping apparatus, in this case 'a closed water-screw pump for peat bog drainage'.

As the industry became more sophisticated and mechanised the old hand-cut systems ended and were replaced by powered apparatus such as this industrial peat excavator.

THE ENCLOSURES AND FUEL ALLOTMENTS

COMMON LAND ENCLOSURES in England, especially the Parliamentary Enclosures, brought major agricultural improvements – but at a cost. The effects on the poor and needy were dramatic, with a significant impact on ecology and landscape too. When commons were enclosed and traditional rights lost there were attempts to compensate cottagers and the landless poor. This was often in small plots of land held in trust for the benefit of the poor as 'poor allotments'. It was important that allotments were a commuted right as compensation for the loss of common and not a gift of charity. Around settlements the allotments were often called 'town lands', 'poor lands', or 'fuel allotments'. Common rights of the medieval manorial system fell into two broad categories, those appendant (related to land ownership) and those appurtenant (related to ancient cottages). These gave rights of common on the village 'waste', including pasturage, piscary (fishing), estover and firebote (wood gathering for house repair and for domestic fuel), and turbary (cutting of peat and turf). Other rights covered cutting of furze for fuel and mineral or stone extraction. There were additional rights granted of 'common in gross' vested in individuals through a grant or by prescription (twenty years of unchallenged use).

For fuels key factors were considered like types of fuels available and accessibility, calorific value and quality (i.e. how much heat they produce and for how long), and costs in effort and energy expended to gather and process. Wood, gorse or furze, fern or bracken, ling or heather, and even cow dung were all collected along with peat or turf. The cost of buying fuel rather than the effort of gathering it was important and variable. In Victorian Berkshire it was estimated a family could cut a year's fuelwood in one week but to purchase the same would take a tenth of a labourer's annual income.

Coal might be cheaper but, depending on quality and the type of fireplace, it was not always favoured. Simple hearths, even with a chimney (which most but not all cottages had by the eighteenth and nineteenth centuries), did not burn coal effectively and generated noxious fumes. Only with metal grates and tall narrow chimneys was coal the better fuel, even though it has twice the calories of wood. But wood was a cleaner fuel to handle and burn and had the

This Victorian illustration shows a wheeled peat cart in Langstroth Dale, Yorkshire.

advantage that its ashes could be sold to industry or agriculture. Peat and turf burn long and slow but produce the traditional 'peat reek'. Furze was commonly used by the poor and similar to wood but quick, hot and short-lived.

There are many examples of varying types of fuel allotments in England. Two which illustrate the importance of peat cutting for fuel are to be found at Holme Moss in West Yorkshire, and the Stockland Turbaries in East Devon.

THE GRAVESHIP OF HOLME, WEST YORKSHIRE

This is England's last community-organised peat cut, others in North Yorkshire and Cumbria having been already quashed. The Graveship, established at the time of Henry V, is overseen by the Constable and his assistant. Each year those owning a hearth in Holme Township can register rights to cut peat and a small area of turbary is allotted. Grazing rights are controlled by the Constable of the Graveship, and fuel rights are for individual domestic use and not commercial exploitation. In times past, rental from the allotments was given to the parish poor (£20 in 1887). By the late 1990s, the rent of £300 per year was available for road maintenance. Around a hundred people still registered

The Graveship of Holme as pictured here near Holme Moss above Holme village in West Yorkshire: England's last community turbary.

their rights in the late 1990s, with only three (Isle of Skye, Hades Peat Pit, and Harden Peat Pits) of the original five turbaries still in use, totalling around 22 hectares. Rights are claimed at the end of Easter with cutting in May and June. In recent years the practice may have ceased.

THE EAST DEVON TURBARIES

The Stockland Turbaries in East Devon near Honiton are maintained by a local project as nature conservation areas. The Turbaries were established by an Act of Parliament when local commons were enclosed in the early 1800s; the enactment was probably made in 1807 with enclosure following somewhat later. The map accompanying the award was dated 1824. The conditions of the Turbaries included that income generated from the site should be put towards '... any other pious and charitable use or uses within the said parish as the feoffees for the time being should think fit'. Collection and distribution were overseen by appointed trustees. Total annual income in 1908 was just over £44, of which £4 went to the parish coal fund, given away at Christmas. In addition to the coal fund there is mention of turbary lands: 'By the Inclosure Award already mentioned there were allotted and awarded to the churchwardens and overseers of the poor of the parish of Stockland for the time being, and their successors for ever the twelve several parcels of land hereinafter described...' The document then goes on to give details of individual turbaries: Bucehays Turbary, Quantock Turbary, Short Moor Turbary, Huntshays Turbary, Hamer Hill Turbary, and Shore Turbary. Memory of their use as working turbaries, abandoned some time between 1910 and 1930, has long disappeared, a photograph of turf cutting in 1914 showing some of the last here. Loss of rural labour in the war effort, the Depression, improved road and rail transport, and availability of coal and coal-burning stoves led to their abandonment.

A peat cart being loaded at Shapwick in Somerset. The rather stout lady seems very precariously balanced on top of the cart wheel, and the ruckles are visible in the middle distance.

OTHER USES OF PEAT

PEAT OR MOOR BATHS

MANY PEOPLE know of peat as fuel, and for horticulture; its other uses are hardly known but wonderfully diverse and interesting. Peat has always been a special resource with almost magical and mystical powers. Peat bogs have the power to preserve and embalm bodies and other items accidentally lost or deliberately placed in them. From early times they were viewed with horror, fascination and awe, and to the present day peat has a reputation for healing properties. It was widely used in European health spas, between the mid nineteenth and the mid twentieth century being promoted as a healing and therapeutic treatment. In Britain this was limited to baths at Buxton, Harrogate and Strathpeffer. Bathing in hot peat was good for the body and perhaps the soul. In Germany, peat rich in radon was especially highly regarded. Today you can purchase peat creams and packs to apply externally and to rub into skin affected by various ailments. It is suggested that the health-giving properties of peat derive from their rich chemistry and biochemistry with trace elements and complex organic compounds. These were exploited in products such as 'sphagnol' soap with its special curative effects helpful for skin complaints. The extracted minerals are marketed both in North America and in Germany as 'Mineral Life' liquid.

PEAT PAPER

Peat paper was made in Ireland from around 1835 and in 1903 there was a factory in Celbridge, County Kildare, that manufactured brown wrapping paper and card from peat moss fibre and waste paper pulp. By 1905 the Callendar Paper Company produced 10 tons of peat paper a week, but this was uneconomical and could not compete with wood or straw paper. Postcards from Irish peat paper were popular around 1900, sent across the world as a link to home for expatriate Irish, or a romantic touch of 'Irishness' for tourists. They were especially popular at Christmas as touching reminders of home.

Opposite: Bird's-eye view of a generalised layout of a peat cutting plant on drained bog.

Top: Digging peat for the peat baths at Buxton Spa, one of the only Peak District images. It is unusual because it shows winter digging.

Middle: Peat spa advertisement from Germany (1936).

Bottom: Peat postcards provided both a romantic view of Ireland for the tourist to send home, and a reminder of the mother country for expatriate Irish. This is 'a bit of old Ireland' on Irish peat paper.

PEAT IN BUILDINGS

In nineteenth-century Europe there were experiments in making peat building blocks for houses with compressed peat and slaked lime, and for under-floor and roof insulation. It was made into peat-and-tar roofing, and pavements. Turf or sod was often used for making poor cottages, because its unique properties made it suitable for rough dwellings in cold and wet conditions. Turf, especially heather turf with the roots intact, was often used to roof poor houses. These materials were free and easy to maintain, dealt with rainfall very effectively, and were very warm. Fenland timber, wattle and daub cottages used turves with reeds for insulation; a cottage in Wicken demolished in the 1950s still had perfect solid turves in its walls. Homes of wet peat or sod were common in more remote or poorer regions, especially Ireland, made mostly with cut turves but sometimes with walls of standing uncut peat. This was most famously used in the peat houses of Flanders Moss near Stirling in the eighteenth century, but also commonly in Ireland. The evidence of such buildings was transient as were the dwellings themselves. They were prone to sinking back into the bog from which they were formed and were also, for obvious reasons, prone to fire damage. Some of the buildings were partly subterranean; others were built of turf stacked on the bog surface. Remarkably, some in Leck near Ballymoney in Ireland were still occupied in the 1950s.

Top: Peat litter bale to supply horses.

Above: This shows a commercial 'moss litter bale' of the type used to supply the army, especially in the First World War, and also all major businesses in towns and cities where haulage by horses was required.

PEAT AND AGRICULTURE

Peat was used extensively for agricultural purposes with vast quantities of peat and shallow turf heaped and burnt to produce peat ash fertiliser. This was sometimes part of the process of 'land improvement' undertaken as 'paring and burning'. In other cases such as the lowland valleys around Reading, peat was extracted from nearby fens, taken to specific 'pits' and burnt to produce ash.

One of the main uses for peat in the late nineteenth and early twentieth centuries was as litter for animal bedding. With huge numbers of animals

Whilst most peat paper cards have novelty but print rather poorly, this high-quality colour postcard on Irish peat paper shows the potential for good results.

powering farms – and in towns and cities there was a big demand for material to keep things clean – peat was ideal. Once soiled, its nutritive qualities enhanced, it went on the land as fertiliser. At Thorne Moors in South Yorkshire, the English Moss Litter Company extracted peat moss up to the 1960s, and from 1923 to 1962 the Midland Litter Company took moss from Fenn's Moss near Wrexham. Although raw peat was widely used as litter by farmers and peasants, wider usage took off in the early 1900s. Processed and packaged as a commercial product during the First World War, with its absorbent and fibrous texture combined with antiseptic properties, it was used extensively as horse bedding for the military.

It was also used as animal feed, mixed with green fodder and perhaps molasses. Again its antibacterial properties may have had a therapeutic effect and it was used either coarse or as a powder mixed into a cattle cake.

PEAT FOR HORTICULTURE

From the 1930s onwards, demand for horticultural peat supplied by hand cutting grew rapidly. However, by the 1950s and 1960s, there was increasing interest in commercial industrial exploitation by machine. This raised operations to another level and sites had to be drier to support big machines and narrow-gauge railways to extract peat in bulk. This meant big drains, huge pumps, landscapes desiccated and ecology devastated. In Ireland not only was there a demand for horticultural peat, but a massive expansion in industrial extraction of peat to fuel power stations. Sites that had evolved slowly over millennia, and their complex interactions with local communities, were destroyed overnight.

A hand-powered moss litter machine.

INDUSTRIAL EXPLOITATION

Since the beginnings of early industry peat and turf have been used in many different ways. These include obvious uses in the agricultural industry, but also in metal smelting, and with distillation treatments to extract valuable organic chemicals. From very early it was realised that peat (like wood) could be distilled to produce chemicals such as coal and oil. Peat was used in primary industries such as the smelting and extraction of metals, and in the subsequent factoring of tools and objects. Fenland blacksmiths used peat rather than wood, mineral coal or charcoal and in Wicken in the 1930s Bill Redit was the last to use turf burnt with wood. He built fires outside close to the 'tyring platform'; the hot metal 'tyre' was swiftly drawn from the fire, hammered onto the wooden wheel, and tightened by cold water shrink-fitting metal to wood.

In Ireland there were many examples of industrial peat use from fuel to chemical extraction. For example between 1906 and 1922, using 8,000 tonnes of hand-cut peat per year, the Marconi trans Atlantic wireless station at Clifden was powered by turf-fuelled steam engines. Peat gas was another product and powered the Robb Linen Factory at Portadown, taking 3,500 tonnes of peat each year.

CHEMICAL EXTRACTION

Peat is a potential source of complex organic compounds much sought after by early industry before petrochemicals. There was considerable interest in processes and industrial works to distil peat in ways similar to charcoaling wood, to produce volatile oil-like products. Industrialists in the pre-petrochemical era saw peat as a source of volatile organic chemicals such as tars. There was already an extensive industry to distil chemicals from wood in processes linked to charcoal manufacture, so the vast reservoir of peat

Below left: Cross-section through an industrial peat gas generator.

Below: An electrically powered peat extraction machine.

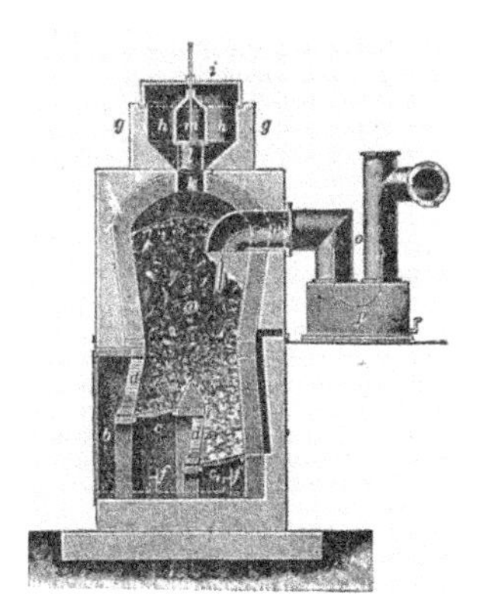

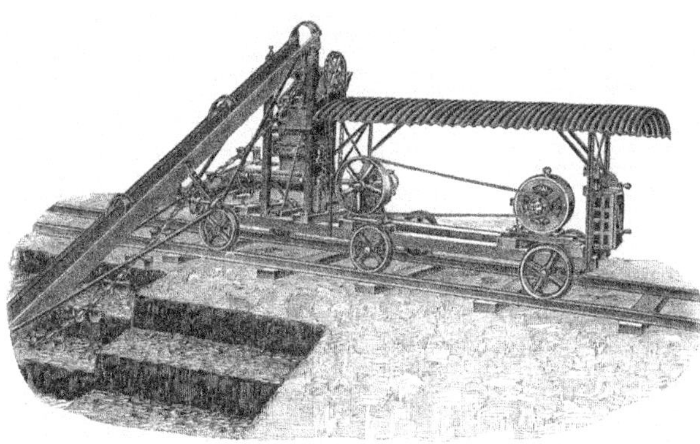

This photograph demonstrates the massive scale of industrial peat extraction or peat winning by the early twentieth century.

seemed an obvious target for experimental industrialists. In the extensive bogs of Germany and Poland chemical engineers sought novel ways to exploit this potential. Peat was used to make dyes and tannins, for candles and fertiliser, and for alcohol and ammonia. These processes were copied in Britain with peat chemical works on Dartmoor (producing 'naptha' and gas), in Sheffield's Ringinglow Bog, around Thorne Moors in South Yorkshire, by the Bettisford Trust Company (taking black peat '... to extract paraffinoid and other chemical products') at Fenn's Moss in 1923, and in the Scottish Islands. None of these lasted very long, the processes being costly and inefficient, and of course hugely destructive of the peat bogs. Many companies lodged patents for specific processes, for example the Bettisford Trust Company in 1924 patented a process of 'destructive distillation'. This used high temperatures to generate benzenes, and lower temperatures for paraffins. The Lewis Chemical Works producing peat oil in the mid nineteenth century was a famous example from the Scottish Isles.

The huge structure photographed here was a peat gas generator in Manchester.

In the late nineteenth century and early twentieth century, carbonising peat processes were developed to extract chemicals. These included Wieland's carboniser in Oldenburg, which with three ovens produced 10 tons of

With the advent of industrial peat cutting in the late nineteenth century came big machines like this industrial peat spreader.

charcoal every day. The peat was broken down into the following constituents: turf charcoal (30 per cent); tar (2.5 per cent); tar water (32.5 per cent); gas and losses (35 per cent). From these products other chemicals could be manufactured with the tar fraction producing lighting oil, creosote, paraffin wax, pitch, methyl alcohol, and calcium acetate. Ammonia and acetic acid were also produced. In the years before effective extraction of coal tar from mineral coal this destructive distillation of peat was very exciting.

Peat gas was also produced in Europe, providing heat and light, though it was not commercially viable. It was used in Ireland to light a number of small towns such as Mullingar in the 1850s, and again during the emergency efforts of the Second World War.

This shows a horse-drawn peat wagon sitting on railway trucks, for carrier transport away from the cutting area.

Railway wagon for carrying peat.

PEAT CHARCOAL

The problem with peat fuel is that it produces little energy for a large volume; it is also bulky and friable, making it difficult to transport from sites of production to its place of use or sale. Turning it into peat charcoal both reduced volume and increased thermal capacity. Charred peat (peat combusted under conditions of carefully restricted oxygen availability) is particularly suitable for metallurgical processes, and also for the production of hydrogen disulphide and activated carbon. It has a high reaction potential due to its great pore surface area, low sulphur content, and low ash content. Activated carbon made from peat has more than 1,000 square metres of surface area per gram. It also has the advantage of being very light with 400 litres weighing only 100 kg. From the 1600s onwards, scientists across Europe experimented with ways to char peat to make useful products. With severe shortages of wood charcoal for metal smelting, there were debates about the potential of peat-based charcoal. There were great successes but also failures; there were severe problems in achieving an even burn and the friability of peat charcoal caused it to be blown away easily in processes where a strong updraught was important. However, carefully prepared briquettes

Railway trucks for peat.

of peat charcoal proved superior to wood charcoal. There were numerous attempts to produce smelted metal using peat charcoal or even raw peat and, whilst some were for a while successful, most failed. The incentives were availability and cost and peat charcoal was used extensively for copper and tin smelting in south-west England. The cause of failure was usually the unpredictable nature of the burn and consequent unacceptable variation in product quality. The smelter at Silverdale in North Lancashire for example used local peat, but the resulting iron often had to be re-smelted and its resale value was lower. Furthermore, the energy expended in drying and charring and then transporting the peat fuel could exceed the heat produced by its burning. In some cases the charcoal was produced for a specific purpose such as for iron-smelting at Creevelea in Ireland in the mid nineteenth century, where there were bonuses of other by-products: the cost of producing peat charcoal for smelting was more than covered by the value of peat petroleum and tar. Peat charcoal was produced in England, especially on the moors of the south-west, probably linked to local metal smelting and a shortage of easily available wood charcoal.

WHISKY

One of the most famous uses of peat is to fire the kilns in whisky distillation and especially in the process of drying the barley before distillation. This imbues the liquor with a uniquely smoky peat flavour characteristic of the West Coast malts such as Islays. Peat was also associated with illicit distilling in the Scottish Highlands, in North Wales, in Ireland, and even in South Yorkshire. Sustaining long-term exploitation presents serious challenges to the malt whisky industry.

Distilling beverages, legal or illegal, has long been associated with peat cutting and peat bogs. This Irish peat paper postcard shows the making of Irish 'mountain dew' or 'poteen' in Connemara.

CONCLUSIONS

IT WILL BE CLEAR from this short overview that peat has had a long and fascinating history. Also in recent centuries it has been a hugely important resource for communities and even industries across Britain. However, the use of peatlands is ultimately threatened by the extraction and use of peat itself. Given the long time required to generate a peat bog, then the level of use witnessed over the last few centuries is clearly not sustainable. Exploitation during the late twentieth century in Britain was devastatingly destructive; worse still, the impacts in Britain are mirrored across Europe and the rest of the world.

Over many centuries, skills, techniques and implements used in the extraction and processing of peat have been exchanged across Europe. Indeed, in Holland and Flanders, in Germany and Poland, in France, and in Scandinavia, peat was used extensively. The Vikings exported peat-cutting tools around north-west Europe, and the Germans later pioneered industrial chemical extraction. Peatlands and peat are truly international resources. The utilisation, whilst destructive, has left a legacy of implements, culture, landscapes and language related to centuries of exploitation, and this is now itself a threatened part of our heritage. This heritage has also been spread across the world by emigrating Irish, Scots, Welsh and English. Peat-cutting tools, landscapes and folklore are to be found in North America, especially in the eastern USA and in Canada. But they are also scattered as far as the Falklands in the South Atlantic and other remote corners of the planet. As in Britain though, these are often forgotten times and neglected memories.

Peat itself is a fascinating substance and has been an important resource. The peats of the British Isles referred to in this book are typical of the north temperate environment. Yet peats are formed and occur wherever plant materials collect in wetlands and cannot decompose, and they are found widely in suitable tropical and sub-tropical climates. Even here they have been badly damaged. Whilst the theme of this short book is a celebration of the heritage and culture of peat use, there are serious problems. Remarkably these are related to issues right at the core of modern society. The main

Opposite: A Stornoway lady carrying peat and knitting, demonstrating an amazing capacity to carry a bulky and heavy load – back-bending work if not back-breaking.

The Peat Inn near Stirling in Scotland, here with peat stacks and carts, is still a famous hostelry.

An evocative picture of rural life in poor Ireland: an Irish 'hovel' with peat spades and turf stacks.

Presumably husband and wife cutting turf for the home; photographed near Oughterard, Connemara, County Galway, Ireland.

Whilst the former Fylingdales early warning system was the centre of an iconic landscape on the North Yorkshire Moors, unknown to many, the area is also a massively peat cut landscape which supplied local farms, villages and even towns like Scarborough and Pickering with peat fuel.

The rural poor had a hard existence but the interior of an Irish cabin with its turf fire could still be snug and warm.

concern is the important role of these peat landscapes in environmental processes such as water management and the release of carbon dioxide as a causative agent of climate change. It is clear that huge amounts of peat have been removed over the last two hundred years in particular, releasing colossal volumes of carbon dioxide into the atmosphere. Furthermore, in removing the peat, that once great sponge that both held back floodwaters and alleviated droughts, has gone. Finally, in a climate of energy shortage there is a renewed interest in peat fuel and its traditions. Hopefully, any future use will seek to be sustainable.

Above: *Peat Bog in Scotland* by J. M. W. Turner, a classic rendering of a familiar scene.

Right: Turfing in Somerset, slitting the cut turves with a sharp cutting spade and lifting with a fork. This image demonstrates the extensive nature of the operation. The turves are presumably laid out on a drying ground.

REFERENCES AND BIBLIOGRAPHY

Anderson, T. *Life on the Levels: Voices from a Working World*. Birlinn Ltd, Edinburgh, 2006.

Ardron, P. A. *Peat Cutting in Upland Britain, with Special Reference to the Peak District*. Unpublished PhD thesis, University of Sheffield, 1999.

Ardron, P. A., Rotherham, I. D. and Gilbert, O. L. 'Peat-cutting and upland landscapes: case studies from the South Pennines' in *Landscape – Perception, Recognition and Management: Reconciling the Impossible?* Proceedings of the Landscape Conservation Forum Conference, 2–4 April 1996, Sheffield. Landscape Archaeology and Ecology, **3**, 65–69.

Berry, A. Q., Gale, F., Daniels, J. L. and Allmark, W. (eds.) *Fenn's and Whixall Mosses*. Clwyd Archaeology Service, Mold, Clwyd, Wales.

Bingham, R. K. *The Chronicles of Milnthorpe*. Cicerone Press, Milnthorpe, 1987.

Charman, D. *Peatlands and Environmental Change*. John Wiley and Sons, London, 2002.

Day, A. *Turf Village: Peat Diggers of Wicken*. Cambridgeshire Libraries and Information Service, Cambridge, 1985.

Denyer, S. *Traditional Buildings and Life in the Lake District*. Victor Gollancz, London, 1991.

Evans, E. E. *Irish Heritage: The Landscape, the People and their Work*. W. Tempest, Dungalgan Press, Dundalk, 1942.

Evelyn, J. *Silva* (1979 reprint of the fifth edition). Stobart and Son, London, 1729.

Feehan, J. and O'Donovan, G. *The Bogs of Ireland: An Introduction to the Natural, Cultural and Industrial Heritage of Irish Peatlands*. The Environmental Institute, University College, Dublin, 1996.

Fenton, A. 'The Shape of the Past 2' in *Essays in Scottish Ethnology*. John Donald Publishers Ltd, Edinburgh, 1986.

Fitzrandolph, H. E. and Hay, M. D. *The Rural Industries of England and Wales. A Survey made on behalf of the Agricultural Economics Research Institute, Oxford. Volume II: Osier-growing and basketry and some rural factories*. Clarendon Press, Oxford, 1926.

Gailey, A. and Fenton, A. (eds.). *The Spade in Northern and Atlantic Europe*. Ulster Folk Museum and Institute of Irish Studies, Queen's University, Belfast, 1970.

Gambles, R. *The Story of the Lakeland Dales*. Phillimore and Co. Ltd, Chichester, 1997.

Gimingham, C. H. *Ecology of Heathland*. Chapman and Hall, London, 1972.

Grant, I. F. *Highland Folk Ways*. Routledge and Kegan Paul, London, 1961.

The extensive boglands were often the only places of potential refuge for evicted and starving peasants in Ireland. This Irish peat paper postcard shows an Irish bog cabin built by evicted tenants.

Hartley, M. and Ingilby, J. *Life and Tradition in the Yorkshire Dales*. Smith Settle, Otley, 1968.

Hartley, M. and Ingilby, J. *Life and Tradition in the Moorlands of North-East Yorkshire*. Smith Settle, Otley, 1972.

Humphries, C. J. and Shaughnessy, E. *Gorse*. Shire Publications Ltd, Princes Risborough, 1987.

Jenkins, J. G. *Traditional Country Craftsmen*. Routledge and Kegan Paul, London, 1965.

Jenkins, J. G. (ed.). 'Studies in Folk Life' in *Essays in Honour of Iorwerth C. Peate*. Routledge and Kegan Paul, London, 1969.

Kerr, W. A. *Peat and its Products*. Begg, Kennedy and Elder, Glasgow, 1905.

Knox, S. A. *The Making of the Shetland Landscape*. John Donald Publishers Ltd, Edinburgh, 1985.

Lambert, J. M., Jennings, J. N., Smith, C. T., Green, C. and Hutchinson, J. N. *The Making of the Broads. A Reconsideration of their origin in the light of new evidence*. John Murray Ltd, London, 1961.

Mann, S. *Cattle Cotton and Commerce. Life around Gatehouse of Fleet in 1792. A Reprint of the First Statistical Account of Scotland for the Parishes of Anworth and Girthon 1792*. Beechwood Publishing, Cambridge, 1993.

Paulus, C. *Unpublished pages relating to the Manor and Parish of Ecclesall including the Enclosures of the Common and Waste Lands there*. Sheffield, 1927.

Peate, I. C. *Tradition and Folk Life: A Welsh View*. Faber & Faber Ltd, London, 1972.

Peter, D. *In and Around Silverdale. The Story of a North Lancashire Village*. Barry Ayre, Carnforth, 1994.

Pinches, S. 'From common rights to cold charity: enclosure and poor allotments in the eighteenth and nineteenth centuries' in Borsay, A. and

Shetland ladies carrying peat in the standard manner of the north and the west.

Shapely, P. (eds.) *Medicine Charity and Mutual Aid*. Ashgate Publishing Ltd, Aldershot, 2007.

Rackham, O. *The History of the Countryside*. Dent, 1986.

Rogers, J. E. T. *A History of Agriculture and Prices in England. Volumes 1–8*. Clarendon Press, Oxford, 1902.

Rollinson, W. *The Lake District Life and Traditions*. Weidenfeld and Nicolson, London, 1996.

Rotherham, I. D. 'Urban heathlands … their conservation, restoration and reclamation' in *Landscape Contamination and Reclamation*, **3** (2), 1995, 99–100.

Rotherham, I. D. 'Peat cutters and their landscapes: fundamental change in a fragile environment' in *Landscape Archaeology and Ecology*, **4**, 1999, 28–51.

Rotherham, I. D. (ed.). *Peatland Ecology and Archaeology: management of a cultural landscape*. Wildtrack Publishing, Sheffield, 1999.

Rotherham, I. D. 'Fuel and landscape – exploitation, environment, crisis and continuum' in *Landscape Archaeology and Ecology*, **5**, 2005, 65–81.

Smith, D. (ed.). *The Third Statistical Account of Scotland, XXIX. The County of Kincardine*. Scottish Academic Press, Edinburgh, March 1951.

Sutherland, P. and Nicolson, A. *Wetland: Life in the Somerset Levels*. Michael Joseph, London, 1986.

Thompson, F. *Harris and Lewis: Outer Hebrides*. David and Charles, Newton Abbot, 1968.

Webb, N. *Heathlands*. Collins, London, 1986.

Withington, L. (ed.). *Elizabethan England*. Scott, London, 1899.

Wright, L. *Home Fires Burning: The History of Domestic Heating and Cooking*. Routledge, 1964.

Young, A. *General View of the Agriculture of the County of Norfolk*. The Board of Agriculture, London, 1804.

INDEX